Texan Crescent
brownish-black above; white spot bands; hindwing below light brown with central pale band

1.25–1.9"

Great Purple Hairstreak
black with iridescent blue; hindwing below dull black with red spots and two tails; abdomen orange-red

1.25–2"

Bordered Patch
wings above black with white spots and an orange band; hindwing below black with orange patch and cream spots along margin

1.4–2"

Polka-Dot Wasp Moth
elongated wings and body; black wings with shiny blue sheen and white spots; shiny blue abdomen with white spots and an orange tip; slender white-tipped antennae; day flying

1.7–2.2"

Red Admiral
forewing with central reddish band, white apical spots, hindwings with broad red border

1.8–2.5"

Baltimore Checkerspot
rows of small white spots; large red-orange marginal spots; seen in wetlands; declining across region

1.8–2.75"

Eastern Buckmoth
wings black with wide central white band with a black eyespot; dark gray abdomen with an orange tip

2–2.9"

Ilia Underwing Moth
dark gray bark-like forewing mottled with brown, black, and cream markings; hindwing black with orange bands and light margin

2.5–3.25"

Zebra Heliconian
Elongated black wings with white or cream stripes; long, narrow abdomen; long-lived; larvae feed on passionflower vines

2.8–4"

Red Spotted Purple
forewing velvety black, hindwing with iridescent blue scaling; mimics toxic Pipevine Swallowtail

2.8–4.25"

male

3–3.75"

Promethea Silkmoth
Wings black with light tan margins; flies during the day

3–4.2"

Mourning Cloak
velvety black; broad irregular yellow wing borders, outer row of purple-blue spots; adults overwinter; young larvae gregarious

male

3–5"

Pipevine Swallowtail
overall black, males have iridescent greenish-blue hindwings; fast, low flight; distasteful to many predators

male

3.25–5"

Black Swallowtail
black with yellow spot bands, black-centered eyespot near tail; female with less yellow; common garden butterfly

3.5–4.5"

Polydamus Swallowtail
black with yellow spot band along wing margin; lacks tails; fast, erratic flight; distasteful to many predators

male

3.5–5"

Spicebush Swallowtail
pale green spots along wing margins, hindwing with extensive greenish-blue scaling, orange eyespot and tail

female

3.5–5.25"

Diana Fritillary

large; black with iridescent blue spots on outer half; mimics toxic Pipevine Swallowtail

female

4–5.8"

Eastern Tiger Swallowtail

female dark form is mostly black and may have faint darker black stripes; hindwing with blue scaling and tail; mimics toxic Pipevine Swallowtail

4.5–5.25"

Palamedes Swallowtail

brownish-black with cream spot bands; yellow streak in hindwing tail; hindwing below with yellow line near abdomen

4.5–6"

Black Witch

large; forewing somewhat pointed; wings dark brown with black, cream and often violet markings; bark-like appearance

0.7–1"

Cassius Blue

lavender blue above; wings below whitish with numerous darker bands and spots; hindwing eyespot rimmed in orange

1.75–2.25"

Ceraunus Blue

male blue above; female brown with blue at base; gray hindwing with an orange, rimmed eyespot and dark spots along edge

male

0.8–1"

Eastern Tailed-Blue

males blue above; females brownish-gray above; hindwing with orange-capped black spot; single hindwing tail; often common to abundant in many areas

0.8–1"

Spring Azure

powdery blue above; female lighter with wider dark margins; gray below with darker gray scaling and increased dark markings

0.8–1.1"

Marine Blue

wings above pale lavender; wings below gray brown striped with white and dark bands

0.9–1.1"

Summer Azure

powdery blue above; female lighter with wider dark margins; chalky gray below with faint dark spots

Western Pygmy Blue

wings brown above with pale blue bases; wings below gray basally and brown on outer half; hindwing with row of small dark spots along outer margin; smallest US butterfly

0.5–.75"

0.8–1.25"

Red-Banded Hairstreak

wings above brown with some light blue scaling; broad red band with a white edge above it; larvae feed on dead leaves and plant material below their host

0.8–1.6"

Tawny-Edged Skipper

forewing brown with orange on leading edge; hindwing below olive-brown; forewing below with orange scaling along leading edge

0.9–1.2"

Henry's Elfin

hindwing with two-toned pattern, gray frosting along margin, and short tail

0.9–1.5"

Banded Hairstreak

band of dark dashes edged in white; hair-like tails resemble antennae and help deflect predators

0.9–1.5"

Oak Hairstreak

hindwing with two tails; two forms throughout South: Florida and coastal Georgia hairstreaks have longer hindwing tails, more red on margin, and a white spot near base

1–1.3"

Eufala Skipper

forewing below gray brown with three pale spots toward apex; hindwing below dull brown with occasional faint spot band

female

1–1.3"

Io Moth

forewing mottled reddish-brown; hindwing reddish-brown with golden bands and a central black eyespot

Northern Broken-Dash

hindwing below brownish with occasional purplish sheen and curved central pale band

1–1.5"

Eastern Tent Caterpillar Moth

forewing brown with two pale bands; hindwing unmarked brown

1–1.7"

Little Glassywing

forewing above dark brown with glassy spots; hindwing below dark brown with purplish sheen and faint central band

1.1–1.5"

White M Hairstreak

iridescent blue above; hindwing below with white band forming an "M" near red patch

1.1–1.65"

Carolina Satyr

wings unmasked brown above; hindwing below with two central dark lines and several small yellow-rimmed dark eyspots along margin

1.2–1.4"

Southern Cloudywing

forewing with prominent band of aligned glassy spots

1.2–1.85"

Northern Cloudywing

forewing with small misaligned glassy white spots

1.2–1.8"

Dun Skipper

forewing unmarked dark brown in males, dark brown with few small white spots in female; hindwing below dark brown often with faint central spot band

1.25–1.4"

female

1.25–1.6"

Sachem
elongated wings; hindwing below golden brown with angled pale spot band or patch

female

1.25–1.6"

Zabulon Skipper
forewing above dark brown with pale spots; hindwing below dark purplish-brown with frosted scaling along outer margin and a white edge along the top margin

1.25–1.75"

Clouded Skipper
forewing above dark brown with small glassy spots; hindwing below dark brown with violet scaling

1.25–1.75"

Horace's Duskywing
forewing brown with small glassy spots in males; female with larger glassy spots and darker brown spots; limited gray scaling

1.25–2"

Juvenal's Duskywing
forewing with extensive gray scaling and small glassy cell spots; common to abundant in early spring

1.25–1.85"

Gemmed Satyr
wings above unmarked brown; hindwing below brown with lighter patch along the margin with blackish-silver spots

1.3–1.7"

Twin-spotted Skipper
forewing above dark brown with white spots; hindwing dark brown with three white spots

1.3–2"

Ocola Skipper
long wings; forewing extends past hindwing; forewing above with small pale spots; hindwing below dark brown with a faint spot band

1.4–2.2"

Hackberry Emperor

triangular wings, forewing apex black with white spots and single black eyespot; males regularly fly out to investigate passersby

1.4–2.2"

Tawny Emperor

triangular wings, more tawny than Hackberry Emperor; forewing apex black without white spots; female much larger than male

1.4–2.25"

American Snout

forewing with orange scaling and apex squared off; gets its name from its snout-like labial palpi; adults have quick, bouncy flight

1.4–1.7"

Little Wood Satyr

two dark lines through wings, each wing with two large yellow-rimmed eyespots; adults perch with wings partially open

1.4–1.8"

Funereal Duskywing

forewing above dark brown with lighter patch toward the apex; hindwing above dark brown with white band along the margin

1.5–2.2"

White-Striped Longtail

wings above dark brown with glassy spots on forewing; hindwing below dark brown with white band through center; hindwing with long narrow tail; southern TX only

1.6–2.3"

Hummingbird Clearwing

wings elongated and reddish-brown with transparent patches; hairy body; olive colored thorax; dark abdomen; day-flying moth

1.7–2"

Hoary Edge

forewing above with glassy golden spots; hindwing with hoary white marginal patch

Common Buckeye
conspicuous eyespots; forewing with broad white band; often perches on bare ground

Brazilian Skipper
long wings; wings above dark brown with large translucent spots; hindwing below red-brown with spots in the center

Long-Tailed Skipper
wings above brown with iridescent green basally; hindwing with long tails

Dorantes Longtail
wings above brown with no green iridescence; hindwing with long tails

Virginia Creeper Sphinx
wings and body elongated; forewing brown with mauve and gray mottling; hindwing orange with brown margin

Silver-Spotted Skipper
elongated forewing; hindwing with large white central patch; showiest skipper in region

Yucca Giant-Skipper
large forewing above elongated and dark brown with yellow spots; hindwing brown with yellow band along the margin, mostly gray-brown below

Common Wood Nymph
forewing with two large eyespots inside broad cream or yellow patch; hindwing yellow with bark-like lines and a number of small eyespots

2.2–2.7"

Southern Pearly-Eye

light brown with darker brown lines; orange-rimmed black eye-spots; orange clubs on antennae; rests with wings closed

2.5–3.2"

Tersa Sphinx Moth

wings and body elongated; forewing brown with bark-like markings; hindwing darker brown with light triangular spot band

2.6–3.5"

White-lined Sphinx

wings and body elongated; forewing brown with cream veins and a cream stripe; hindwing brown with pink band; feeds like a hummingbird at flowers

4–5.8"

Polyphemus Moth

wings above warm tan to light brown; hindwing above with large black eyespot marked with yellow and blue; hindwing below mottled brown

4.4–5.8"

Cecropia Silkmoth

wings black-brown with pale margins, a red and white band, and a prominent light central crescent-shaped spot; body pattered red and white

4.5–6"

Giant Swallowtail

broad, intersecting yellow spot bands; hindwing tail with yellow center; larvae resemble bird droppings and feed on cultivated citrus

Mostly gray

Gray Hairstreak
0.9–1.4"

wings uniformly gray with narrow black line edged in white; hindwing with orange-capped black spot near tail; arguably most common hairstreak in region

Pink-spotted Hawkmoth
3.8–4.75"

wings and body elongated; hindwing gray mottled with brown and white, black and white bands, and a pink base; abdomen gray with pink-and-black spots

Mostly green

Juniper Hairstreak
0.9–1.25"

olive green with white spot band; always found near stands of Eastern Redcedar trees

Malachite
3–4"

wings above dark brown with large green spots; wings below light tawny orange with large green spots and white bands giving an overall agate-like appearance

Luna Moth
3–4.5"

large; wings pale green; hindwing above with single eyespot in center and long curved tail; fuzzy white body; males have fern-like antennae

Pandorus Sphinx
3.25–4.4"

wings and body elongated; wings mottled green, olive, and often brown with some pink along the veins

Mostly orange

0.6–1"

0.65–0.85"

Little Metalmark
wings above reddish-orange with
numerous metallic spots and bands;
wings below lighter orange; often
rests on the undersides of leaves

Southern Skipperling
tiny; wings elongated; hindwing
with central pale streak; low,
weak flight

0.8–1.25"

0.9–1.1"

American Copper
forewing orange with dark spots
and margins, hindwing grayish
with wavy orange band

Least Skipper
rounded wings above orange
with dark border; hindwing below
unmarked orange; adults have
low, weak flight

1–1.3"

1–1.3"

Harvester
forewing above orange with wide
black borders and spots; hindwing
below reddish-brown with white
outlined spots; only US butterfly
with carnivorous larvae

Peck's Skipper
wings above dark brown with
orange markings; hindwing below
brown with large irregular central
yellow patch

male

1–1.5"

1–1.5"

Whirlabout
hindwing golden with two broken
dark spot bands in male; hindwing
tawny brown with two broken
dark spot bands in female

Fiery Skipper
elongated wings; male orange
above with black margin; female
more brown; hindwing below
gold-orange with some black spots

Southern Broken-Dash
hindwing below golden-brown to tawny orange with occasional purplish sheen and curved central pale band

1–1.5"

Phaon Crescent
orange with black bands, spots, and borders; forewing with cream central band; hindwing below with pale crescent spot on outer margin; hindwing below varies seasonally

1.2–1.6"

Pearl Crescent
orange with black bands, spots, and borders; hindwing below with pale crescent spot along outer margin; widespread and common

1.2–1.75"

Delaware Skipper
wings above golden orange with dark borders and veins; wings below unmarked golden orange

1.2–1.9"

male

Zabulon Skipper
wings above orange with dark brown margins; hindwing below yellow with brown at the base and outer margin

1.25–1.7"

male

Hobomok Skipper
wings above orange with irregular dark brown border; hindwing below dark brown with broad central yellow patch

1.4–1.6"

Sleepy Orange
orange with irregular dark borders; hindwing below butter yellow with brown markings; seasonally variable; looks orange in flight

1.4–2.4"

Silvery Checkerspot
tawny orange with black markings and borders, hindwing with white-centered black spots

1.5–2"

Mostly orange

1.5–3"

Orange Sulphur
bright yellow, hindwing with large red-rimmed silver spot and dark spot band; often in clover and alfalfa fields; appears orange in flight; females occasionally white

1.75–2.25"

American Lady
orange with dark marks and borders, forewing apex squared off; hindwing below with agate pattern and two large eyespots; common in open disturbed sites

1.75–2.4"

Painted Lady
pinkish-orange with dark marks, forewing apex black with white spots

1.8–2.5"

Eastern Comma
forewing apex squared off, irregular, jagged margins; hindwing with stubby tail; hindwing below with silver "comma" in center

2–3.2"

Variegated Fritillary
forewing somewhat elongated; wings with light central band and darker orange basally

2.25–3"

Question Mark
forewing apex squared off with jagged margins; hindwing with tail and lavender border; hindwing with silver question mark-like spot

2.25–3"

Goatweed Leafwing
orange above; forewing with hooked apex; hindwing with tail; wings have dead leaf pattern below; difficult to approach

2.5–3.5"

Viceroy
orange with black veins and borders, hindwing with postmedian black line; resembles monarch; typically found near wetlands

Mostly orange

2.5–3.7"

Large Orange Sulphur
male a bright unmarked orange above; females pinkish-white or light yellow-orange; forewing below with straight central line from the apex

3–3.4"

Gulf Fritillary
elongated wings; orange with black markings and black-outlined white spots on forewing; hindwing below with elongated silver spots; migratory

3–4.5"

Great Spangled Fritillary
bright orange with black lines and spots, wings dark basally; hindwing below with silvery spots

male

3.4–5"

Diana Fritillary
wings unmarked blackish-brown with broad orange margins; largest fritillary in region; males and females dimorphic

3.5–4.5"

Queen
mahogany above with black borders and white forewing spots; hindwing below mahogany with black veins; larvae feed on milkweed

3.5–5"

Monarch
large; orange with black veins and borders, forewing with white apical spots; can migrate thousands of miles; larvae feed on milkweed

4–6.25"

Regal Moth
large; wings elongated; forewing grayish with orange veins and yellow spots; hindwing orange; body orange with yellow markings; females much larger than males

Mostly pink

Bella Moth
forewing yellow with white bands containing black spots; hindwing pink with an irregular black margin; day flying

1.1–1.75"

Rosy Maple Moth
wings bright pink with central yellow band; body fuzzy and yellow

1.5–2.1"

1.6–2"

Virgin Tiger Moth
forewing black with intersecting cream lines; hindwing pink with large black spots; abdomen pink with black spots

Mostly red

female

3–3.75"

Promethea Silkmoth
wings reddish-brown with paler outer half and tan margins; forewing with eyespot toward apex; hindwing with angled central cream spot

Mostly white

Common/White Checkered-Skipper

wings black with scattered white spots; hindwing below white with tan bands

Tropical Checkered-Skipper

wings black with scattered white spots; male with blue-gray hairs basally; hindwing below white with darker bands and spots

1–1.5"

1-1.5"

1-1.6"

Barred Yellow

forewing with dark tip and bar along edge; hindwing below seasonally variable; low, erratic flight

1.3-1.75"

Falcate Orangetip

forewing with hooked apex; forewing with single black eyespot; forewing with orange tip in males and white tip in females; hindwing underside a marbled gray brown

1.5–2.25"

Cabbage White

forewing with black tip and single black eyespot in male; two eyespots in female; accidentally introduced from Europe

1.5–2.4"

Checkered White

white and black checkered pattern; dark scaling is seasonally variable; found in open, disturbed areas

2–2.85"

White Peacock

wings above white with brown scaling, black spots and orange bands; wings below paler; hindwing with short stubby tail

2.5–3.2"

Great Southern White

male white with narrow black forewing margin; female a dirty white to gray; aquamarine antennae tips

Mostly white

Great Leopard Moth
elongated white wings with mix of solid and hollow black spots; white abdomen with orange and metallic-blue spots

Zebra Swallowtail
striped white and black; hindwing with red eyespot and long tail; low, quick flight; uncommon in cities

Mostly yellow

Dainty Sulphur
elongated wings, forewing with black tip; forewing below with orange and a few black spots

Little Yellow
small; yellow to nearly white, several dark spots or patches; seasonally variable; low, erratic flight

Isabella Tiger Moth
elongated, yellow-orange forewing with faint brown markings; hindwing paler with orange or pinkish flush; abdomen orange with black spots

female

Clouded Sulphur
yellow to greenish-yellow below, pink wing fringes, hindwing with large red-rimmed silver spot; common to abundant in clover and alfalfa fields

male

Io Moth
forewing yellow with darker markings; hindwing yellow with orange bands and a large eyespot

Southern Dogface
forewing with pointed apex and black cell spot; named for forewing pattern that resembles the head of a dog in profile; only sulphur with pointed forewings

Mostly yellow

male

2.8–.5"

3.5–5.5"

Cloudless Sulphur

large, greenish-yellow with limited dark spots, hindwing with small central silver spot; largest sulphur in region; migrates south in fall

Eastern Tiger Swallowtail

large, wide black stripes; broad black wing margin, hindwing with single tail

3.5-5.8"

4–6.7"

Two-Tailed Swallowtail

large; wings yellow with bold black stripes and margins; hindwing with two long tails; TX, west OK only

Imperial Moth

large; elongated yellow wings with varying degrees of purplish-brown markings; female is much larger than the male

Larvae (Caterpillars)

NOTE:
Some caterpillars possess stinging hairs or spines and can cause allergic reactions or irritation; before you handle an unidentified caterpillar, put on gloves. (Or better yet, just take pictures.) In the text below, we've called out the species where handling caterpillars may be an issue.

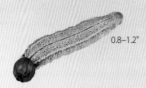
0.8–1.2"

Long-tailed Skipper
green body with yellow stripes and small black spots; brownish-black bulbous head marked with two orange spots

1–1.5"

Fir Tussock Moth
hairy, gray-to-cream body with two long black tufts extending forward past the head and one off the rear; four compact brushy hair tufts on the back; **Caution:** Hairs may cause irritation.

1.3–1.6"

Gulf Fritillary
bright orange body with black spines

1.4–1.75"

Zebra Heliconian
white body with black spots and black spines

1.4-2"

Monarch
body striped with alternating bands of black, white, and yellow; a pair of long, black filaments on both ends

1.5-1.75"

Cloudless Sulphur
green to yellow body with a yellow side stripe, blue side patches, and numerous small black spots

Larvae (Caterpillars)

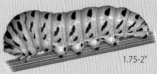

1.5-1.9"

Pipevine Swallowtail
dark brown to reddish-brown body with rows of bright orange bumps (tubercles) and fleshy projections off the sides; resembles a centipede

1.75-2"

Black Swallowtail
green with black bands with yellow-orange spots

1.75–2"

1.75–2.2"

Mourning Cloak
dark gray to black body with numerous small white spots, reddish patches on the back; black spines

Giant Swallowtail
dark brown body with a white to cream-colored saddle and rear end; looks like a bird dropping

1.75–2.5"

1.8-2.2"

Brazilian Skipper
green semi-transparent body with a pale orange head marked with a central dark triangle

Io Moth
bright green body with a white side stripe bordered in red, and clusters of black-tipped green venomous spines; **Caution:** Spines can cause painful stings.

2–2.25"

2.3–3"

Spicebush Swallowtail
green and brown body separated by a yellow lateral stripe; small blue spots; enlarged thorax with two prominent eyespots

Polyphemus Moth
green with yellow diagonal side stripes and small reddish wart-like bumps; head is brown

Larvae (Caterpillars)

3–4"

2.75–3"

Giant Leopard Moth
black body with dense black hairs and reddish spiracles and bands between each segment

Tomato Hornworm
bright green body with seven white stripes on the side; a prominent curved horn off the rear

4–4.5"

4.5–5.5"

Cecropia Moth
bluish-green body with bright blue, red, and yellow tubercles (outgrowths), each with short spines; **Caution:** Tubercles may cause irritation.

Hickory Horned Devil
greenish-blue body with an orange head; long, curved red-and-black horns

JARET C. DANIELS, Ph.D., is a professional nature photographer, author, native plant enthusiast, and entomologist at the University of Florida, specializing in insect ecology and conservation. He has authored numerous scientific papers, popular articles, and books on gardening, wildlife conservation, insects, and butterflies, including butterfly field guides for Florida, Georgia, the Carolinas, Ohio, and Michigan. Jaret currently lives in Gainesville, Florida, with his wife, Stephanie.

Adventure Quick Guides

Only South & Southeast Butterflies & Moths
Organized by color for quick and easy identification

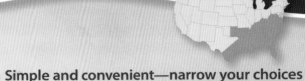

Simple and convenient—narrow your choices by color, and view just a few species at a time

- Pocket-size format—easier than laminated foldouts
- Professional photos showing key markings
- Easy-to-use information for even casual observers
- Size ranges for quick comparison and identification
- The basics of butterfly and moth anatomy

Collect all the *Adventure Quick Guides* for your area

NATURE/BUTTERFLIES/SOUTH & SOUTHEAST

ISBN 978-1-64755-213-8   **U.S. $9.95**

50995

Adventure
PUBLICATIONS
an imprint of AdventureKEEN